Original Title

A great Discovery

The doubling of time

1st Edition: April 2022

Edition :
editorial reconocerse
www.**editorialreconocerse**.com

www.**desdoblamiento**.es

ISBN: 978-84-942181-2-5
Legal deposit: DL T 430-2022

Printed by: Romanyà Valls, S.A

Printed in Spain

Index

Chapter 4

Chapter 5

Chapter 6

Chapter 7

The planetary orbits of the solar system

Chapter 8

The two solar cycles of the doubling of time

A SHORT PREAMBLE

First of all, try to look at the scientific aspect of time doubling which has been the subject of several publications, approved by competent referees.

However, even if I simplify this presentation to the extreme - as I have tried to do at the risk of displeasing the purists - it remains somewhat complicated for those who are eager to quickly learn about its direct applications in everyday life.

It therefore seems useful to me to suggest that you do not dwell too much on the demonstrations but retain the results, the simple logic, the multiple applications and, above all, the conclusions which are as vital as they are fundamental.

Jean Pierre Garnier Malet

INTRODUCTION

Our universe is not the result of chance, and even less the consequence of a mysterious Big Bang which, from nothing, would have made a whole. It is the result of a universal movement, capable of defining and doubling space and time.

As Einstein assumed, before he was forced to "admit his mistake", this doubling looks like a static universe breathe. In fact, it is necessary and sufficient to have an α particle and an omega Ω horizon at the beginning of the doubling.

With perfect scientific logic, this curious "breathing" of the observable universe also animates our solar system that "doubles" our time!

A contraction precedes an expansion that precedes a contraction, according to a first cycle of 12 times 2160 years, related to the equinoxes and a second (Milankovitch cycle) which, according to a law whose importance we shall see, corresponds to 5/3 of the first. At the end of the first cycle, a 13th transition period reveals the existing future potentials, allowing to better prepare the next cycle!

This universal movement doubles and differentiates 3 times (past, present, future) which exchange multiple information with speeds independent of the speed of the source and of the observer.

Used by all particles, these three doubled and simultaneous times are an integral part of our lives!

Driven by this doubling movement, our solar system makes our life on earth an indispensable link between a future - where time is very slow - and a past - where it is very fast - both evolving at the same time as our present - where time seems to us normal and universal.

Therefore, we have the possibility to create the best future potentials there where a very slow time exists which does not appear in our time. At each moment, these potentials can be memorized there where there is such a fast time that our present time does not have time to appear.

This very fast "elsewhere" seems to be an instantaneous past where the best synthesis is made that can then be suggested to us in our present. The fast exchange of information in imperceptible times gives us the impression of a permanent and instantaneous chance.

By ignoring the law of time that animates our solar system, we can experience the worst or the best of futures finding out that we are our own god or our own demon!

Hasn't the obscurantism of the last few centuries led us to mistrust our intuitions, predictions, certainties, clairvoyance or beliefs... with good reason?

In the past, the law of "the division of time" was well known to the "doctors of the law"!

The Greek alphabet was the geometrical and numerical result and the hour (Ωρα in Greek) or the division of time was not the result of chance but the consequence of a very great scientific knowledge...

It is a universal knowledge whose simplicity you are going to discover and that the "predictable" obscurantism of the Middle Ages had tipped over into the dark "oubliettes of time"!

CHAPTER 01

THE DOUBLING

1. Doubling: radial path - tangential path

By rotating at speed ϕ around its diameter, an alpha α circle of radius r generates an alpha particle (α) which thus becomes the horizon of its internal particles (α/2), (α/4), (α/8)... according to the figure below:

THE UNIVERSAL DOUBLING MOVEMENT

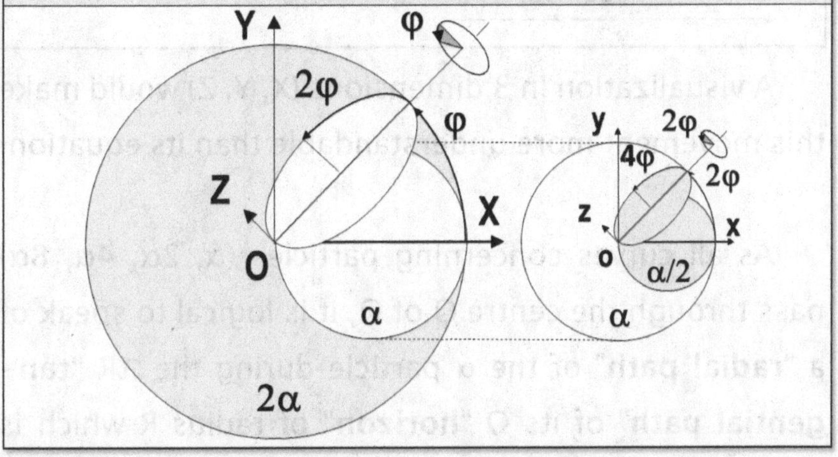

Likewise, the 2α circle rotates on itself at speed φ, as its radius, diameter of the α circle.

The circles α, 2α, 4α, 8α=Ω, or α, α/2, α/4, α/8=Ω/8² make the same movement.

An equation defines the movement of the α particle in the Ω horizon:

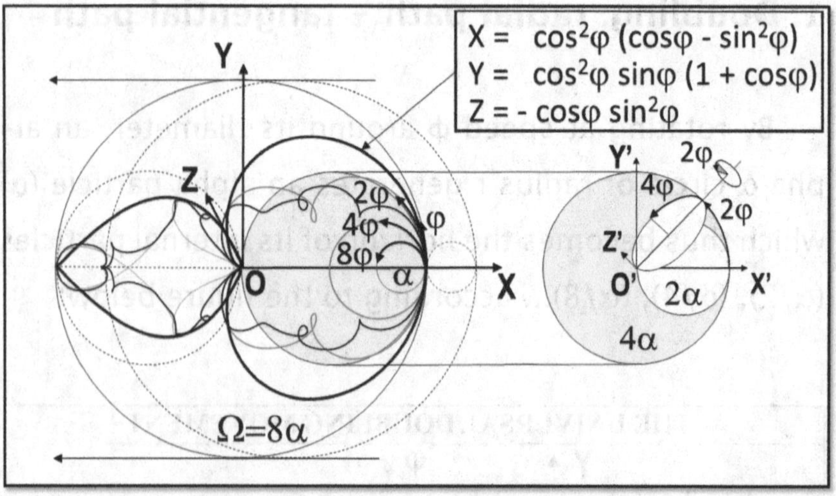

A visualization in 3 dimensions (X, Y, Z) would make this movement more understandable than its equation.

As all curves concerning particles (α, 2α, 4α, 8α) pass through the centre O of Ω, it is logical to speak of a "**radial path**" of the α particle during the πR "**tangential path**" of its Ω "**horizon**" of radius R which is itself a particle in an 8Ω horizon.

A ¼ turn of Ω puts the "**α particle tangential**" in the centre of Ω where it becomes the "**α radial particle**".

A ½ turn (πR=180°) of Ω puts it back tangential on Ω:

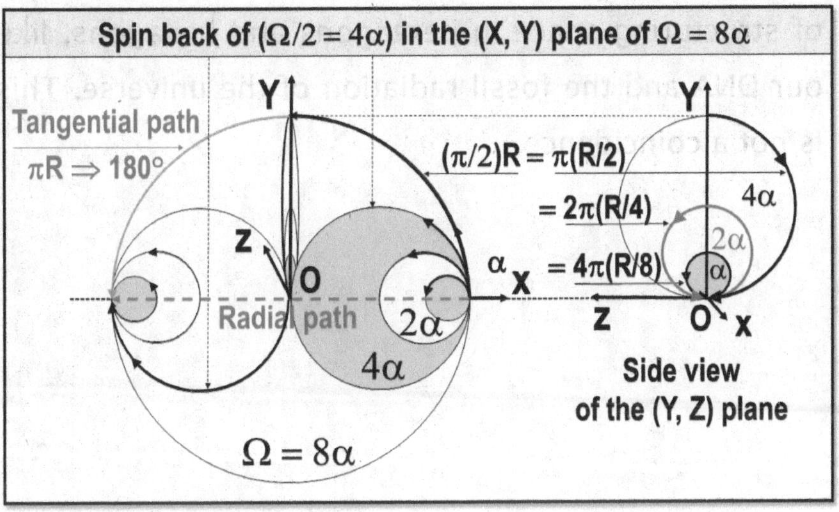

This ½ turn of Ω spins back the (Ω/2 = 4α) plane in the (X, Y) initial plane. In this movement, Ω is therefore **the limit or "the Ω horizon"** of "**the α particle**".

Thus, this movement doubles the α = Ω/8 particle in a "α tangential particle" on the Ω = 8α horizon and a "α radial particle" that passes through the centre of Ω.

This Ω horizon is also a tangential particle on its 8Ω horizon and a radial particle in its 8Ω horizon. This allows us to say that the observer of a particle in its horizon is, therefore and at the same time, the observer of a horizon of particles.

We will see that this movement has the particularity of structuring space in pentagons and hexagons, like our DNA and the fossil radiation of the universe. This is not a coincidence.

CHAPTER 02

(INTERNAL)$^-$ AND (EXTERNAL)$^+$ OBSERVER: 2/3 LAW

2.1 (Internal)$^-$ and (external)$^+$ observer

An observer (particle or group of particles) evolves in a time whose flow he can measure. Thus, we observe a day of 24 hours, an hour of 60 minutes, a minute of 60 seconds...

We shall see that this very ancient division of time is linked to the doubling movement and that between two perceptible times there is always an imperceptible but never zero lapse of time.

Let us imagine:

– A particle or an (external observer)$^+$ who evolves in a time whose flow is much (faster)$^+$ than ours and such that an imperceptible lapse of time for him co-rresponds to 3 days for us.

– A particle or an (internal observer)⁻ who evolves in a time whose flow is much (slower)⁻ than ours and such that an imperceptible lapse of time for us corresponds to 40 days for him.

Through information exchanges between the diffe-rent observers in imperceptible lapses of time, an ins-tantaneous information from an (external observer)⁺ gives us a synthesis of all our future possibilities, ex-perienced at length, in their time, by (internal obser-vers)⁻.

I have called “time openings”, these lapses of im-perceptible times.

2.2. The radial path separates 2 observers:

– In its Ω horizon, **(the internal observer)⁻** observes 8 radial paths (8α) of the α particle.

– In its 8Ω horizon, **(the external observer)⁺** observes at the same time the radial path (Ω/8 = α) of the Ω particle that adds 1/8th to the radial path 2R = 1 of α:

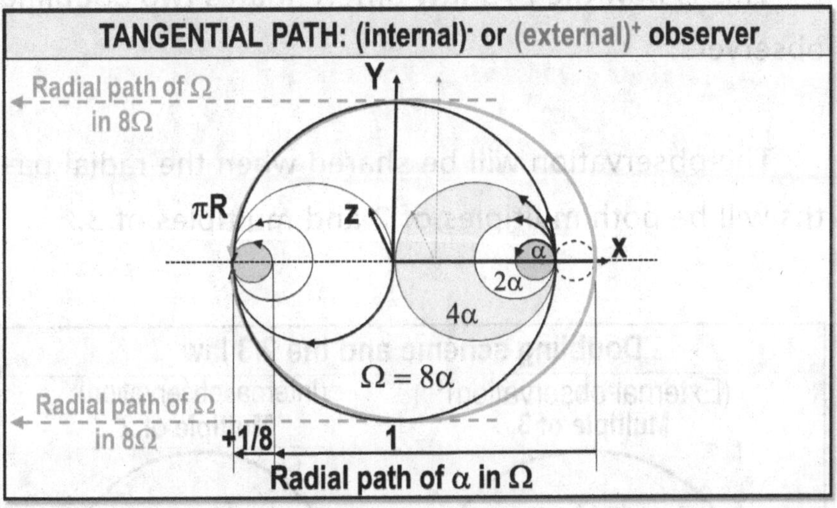

Thus, (the external observer)⁺ observes a 9th radial path that (the internal observer)⁻ cannot observe.

Therefore, (the internal observer)⁻ observes a multiple of 2 radial paths while the (external observer)⁺ observes a multiple of 3 radial paths.

2.3. The 2/3 law

Let us remember that with the doubling of time movement, we are in our time at the same time (the internal observer)$^-$ of an (external observer)$^+$ evolving in a very fast time and (the external observer)$^+$ of an (internal observer)$^-$ evolving in a very slow time.

This is how the **2/3 law** differentiates two doubling observers.

The observation will be shared when the radial paths will be both multiples of 2 and multiples of 3.

Doubling scheme and the 2/3 law

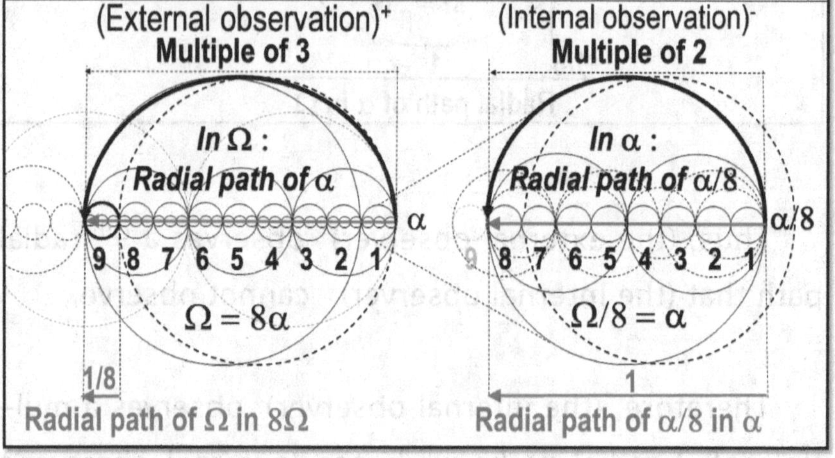

Now, a particle is always a horizon of particles and a horizon, a particle in its horizon.

We can therefore have a juxtaposition of the radial paths of a horizon and a particle when these paths are at the same time multiples of 2 and multiples of 3.

Thus, this 2/3 law is a universal law that we find in our solar system and this is what we will demonstrate.

CHAPTER 03

DILATION AND ACCELERATION OF THE α IN THE Ω REAL TIME AND IMAGINARY TIME

3.1. Dilation and acceleration of the α in the Ω

In the (Y, Z) plane, perpendicular to the (X, Y) and (Z, X) planes, of a 3-dimensional (X, Y, Z) space, the ¼ turn $(\pi/2)R$ of Ω of radius R juxtaposes the α, 2α, 4α particles in the (Y, Z) plane.

The α particle uses this juxtaposition as a “switch” to expand (×2) and accelerate (×2) itself:

Dilation (×2) and acceleration (×2) of the α particle

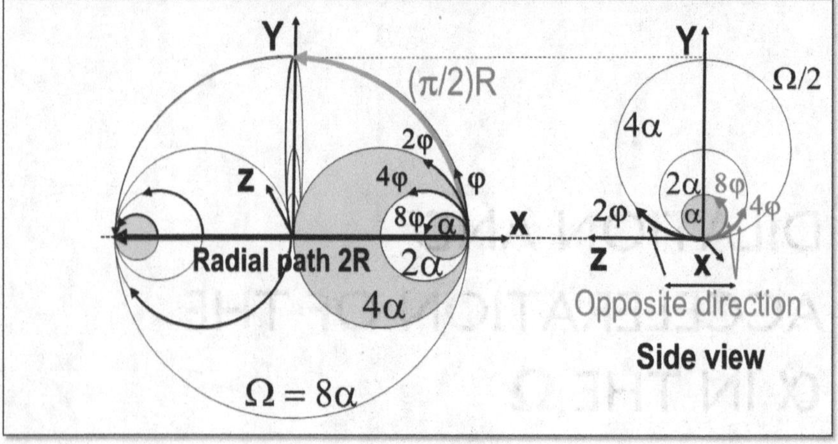

The dilation and acceleration of the α moving particle in the Ω horizon creates an annular space that expands in such a way that the acceleration may not be observable in the observer's time.

Thus we can speak of a circular particle that moves by expanding.

3.2. Dilation $\alpha \rightarrow 2\alpha$

With a dilated 2α particle, the $\pi/2$ rotation in the 2nd (Y, Z) plane is twice faster than the $\pi/4$ rotation in the (X, Y) initial plane:

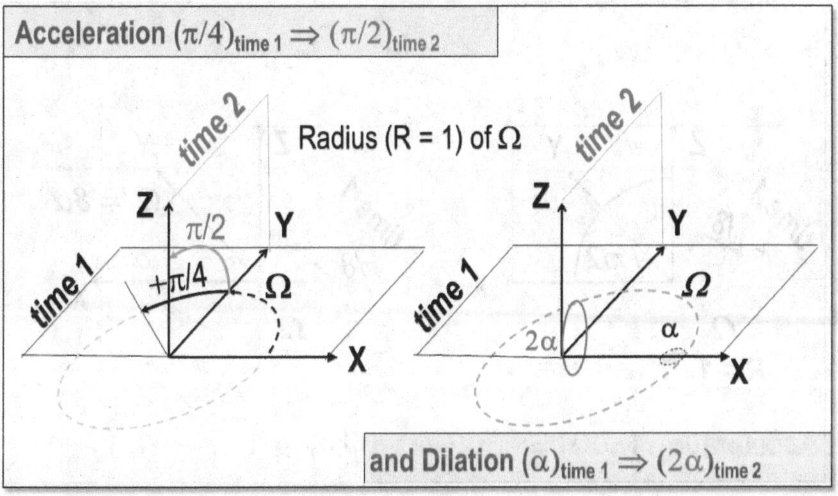

3.3. Dilation 2α → 4α

In the same way, in the 3rd (Z, X) plane a doubling time 3 is 4 times faster than time 1 in the (X, Y) initial plane:

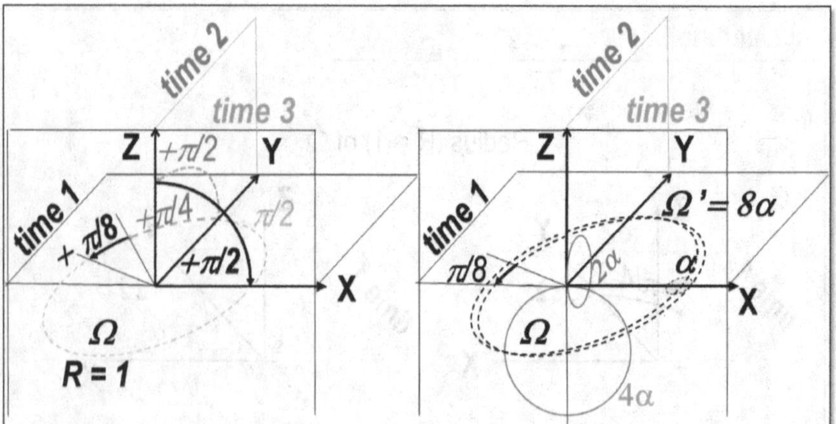

Acceleration $(\pi/8)_1 \Rightarrow (\pi/2)_3$ Dilation $(\alpha)_1 \Rightarrow (4\alpha)_3$

The α initial particle returns to the 1st (X, Y) plane, with a dilation (x8) of α (such as 8α = Ω) and an acceleration (x8) of its movement.

3.4. Dilation 4α → 8α

Thus, in the initial plane, the dilated Ω'= 8α particle doubles the Ω horizon (of radius R=1). This Ω horizon completes its π/8 rotation, during the half turn π of the Ω' dilated particle and the 9th radial path of the α particle in the Ω horizon.

This 9th radial path is imperceptible for the observer who considers Ω as his observation horizon:

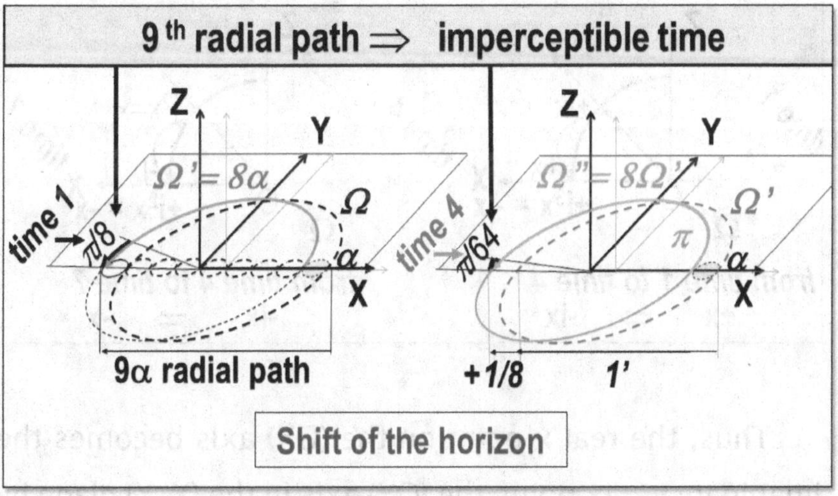

3.5. Imaginary numbers to define rotations:

In mathematics, we know that in addition to the real numbers x, there are imaginary numbers ix such as :

$i^2 = -1$ and $i^0 = 1$.

They are used to define rotations:

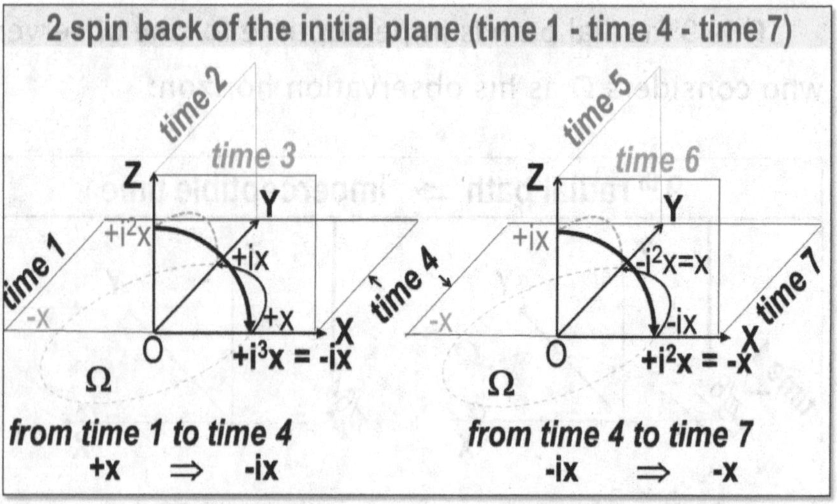

Thus, the real x point on the (OX) axis becomes the imaginary y = ix point the (OY) axis in the (X, Y) plane by a rotation of a quarter of a turn ($\pi/2$), called derivation.

A 2nd quarter turn (or derivation $\pi/2$) in the 2nd (Y, Z) plane, transforms (ix) in $i(ix) = i^2x = -x$ in the 3rd (Z, X) plane.

A 3rd quarter turn (or derivation ($\pi/2$) in this 3rd plane transforms $-x$ in $-ix$ in the (X, Y) initial plane.

Thus, the movement is repeated in this initial plane, but with $-ix$ instead of x.

These rotations, (or "derivations" 8 times faster at time 4 than at time 1) are in an "imaginary negative" plane $-i(X, Y)$, since x has become $-ix$ in the (X, Y) initial plane.

In this $-i(X, Y)$ plane of time 4, a $\pi/2$ rotation (or derivation) transforms $(-ix)$ in $(-i^2x = x)$.

By a $\pi/2$ rotation (or derivation) in the 2nd new (Y, Z) plane of time 5, $(-i^2x = x)$ becomes $(-i^3x = ix)$.

By a $\pi/2$ rotation (or derivation) in the 3rd new (Y, X) plane of time 6, $(-i^3x = ix)$ becomes $(i^2x = -x)$ in the new (X, Y) plane of time 7, superposed to the (X,Y) initial plane.

This rotation of 180° in the (X, Y) initial plane transforms x in $-x$.

In the plane of "time 4", the Ω'= 8α dilated horizon makes the same movement but **in an imperceptible or "imaginary" time** in the initial Ω horizon.

NB: A complex number (z = a + ib) is the sum of a real number a and an imaginary number ib, where b is a real number. We can therefore say that the juxtaposition of the (X, Y) initial plane of time 1 and the (X, Y) plane of time 4 forms a "complex plane" with a real part and an imaginary part. Indeed, the number z defines a point in the initial plane of time 1 and the number $i^4z = z$ defines the same point in the plane of time 4.

3.6. Imperceptible times

It is thus in the (X, Y) initial plane that the movement defines the "**time 4**" of the ($\Omega'=8\alpha$) dilated horizon, 8 times faster than the "**time 1**" of Ω. It also defines the "**time 7**" of the ($\Omega''= 8\Omega'= 64\alpha$) dilated horizon, 8 times faster than "**time 4**":

9th radial path ⇒ imperceptible time

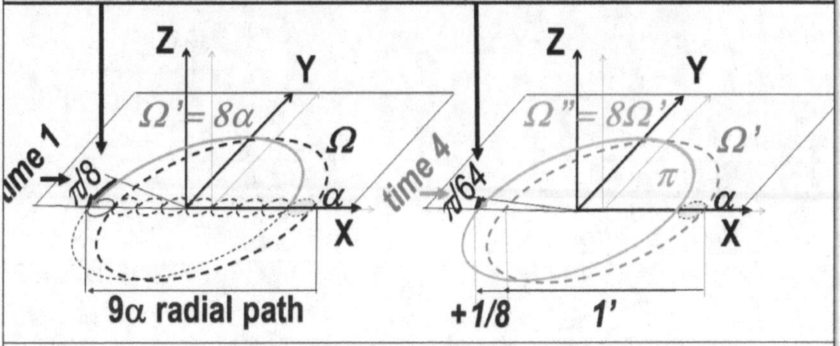

Acceleration and Dilation:

$(\pi/64)_{time\,7} \Rightarrow (\pi/8)_4 \Rightarrow (\pi)_1$ and $(\alpha)_7 \Rightarrow (8\alpha)_4 \Rightarrow (64\alpha)_1$

In the (X, Y) initial plane, the π of Ω tangential path (or 180° rotation) thus corresponds to the $\pi/8$ of Ω' tangential path (or 22.5° rotation).

Conclusion:

Corresponding to a 9th radial path, this rotation can define an imperceptible time for the observer of time1.

CHAPTER 04

EXCHANGE RADIAL – TANGENTIAL (GαrmαA)$^{+}$ HORIZON AND THE PLANETARY SYSTEM

4.1. Exchange radial-tangential in the imperceptible

The rotating movement of "**time 1**" begins on the real side (x) of the (X, Y) initial plane. That of "**time 4**" begins on the imaginary side (-ix) of the same plane. The rotation still corresponds to the "imaginary i number", seen previously and such as $i^2 = -1$.

Indeed, the 9th radial path in the Ω initial horizon is not observable by an (**internal observer**)⁻ because the latter is integral with its Ω horizon whose movement it follows in the 8Ω horizon.

However, it exists for the (**external observer**)⁺ who considers the Ω horizon as a particle in movement in the 8Ω horizon.

We can say that, doubled from the Ω initial horizon, the Ω' horizon appears on the "imaginary" side of the (X, Y) initial plane during this 9th radial path.

This "imaginary" side of the initial plane may therefore be equally imaginary or imperceptible to the (external observer)⁺. In this case, the 9th radial time is not observable by either the (internal observer)⁻ or the (external observer)⁺, but it has an observable influence in the other times.

There is therefore a "**perceptible side**" but also an "**imperceptible side**" of the initial plane that we can imagine, let us say a "real imaginary".

Thus, the 9th imperceptible time of the radial path allows the radial particle in the 1st ($\Omega_1 = 8\alpha_1$) horizon, an exchange with the tangential particle which doubles it on Ω_1.

4.2. The initial horizon or (Gαrmα)$^+$horizon

After the radial–tangential exchange of the 9th time of a Ω initial horizon, the 10th time of Ω is the 1st one of the (Ω_1 =8Ω) horizon but with an acceleration of the movement from 1 to 10:

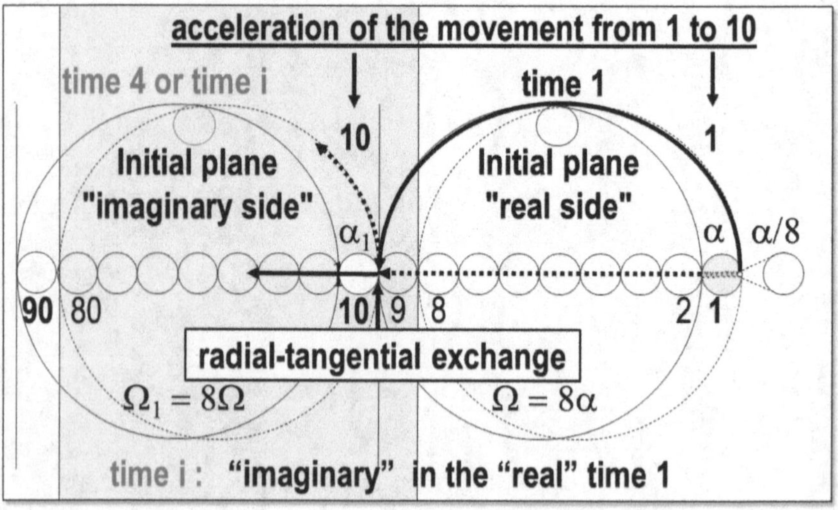

The observer of the 2nd time finds the initial conditions of the observer of the 1st time, after the 8 radial paths of the α_1 particle, but with a $\alpha_2 = 8\alpha_1$ dilation and with an acceleration of the movement from 1 to 10.

The doubling thus begins at the 10th radial path after the 9 made in a Ω=G$^+$ 1st horizon, imperceptible to the doubled observer in the Ω_1 horizon.

As a wink to myself, **I called "Gαrmα" this 1st horizon in our solar system (Garnier Malet).**

Doubled from the initial observer of time 1, the observer of time 2 carries out in an accelerated time the future of this initial observer for whom this future is an imperceptible potential in an "imaginary" plane.

It will be the same for the observer of time 3,... then of time 4. For, after the "Gαrmα" or Ω=G+ horizon the doubling continues with Mercury Ω_1, Venus Ω_2, the Earth Ω_3, Mars Ω_4, and the Asteroid belt Ω_5:

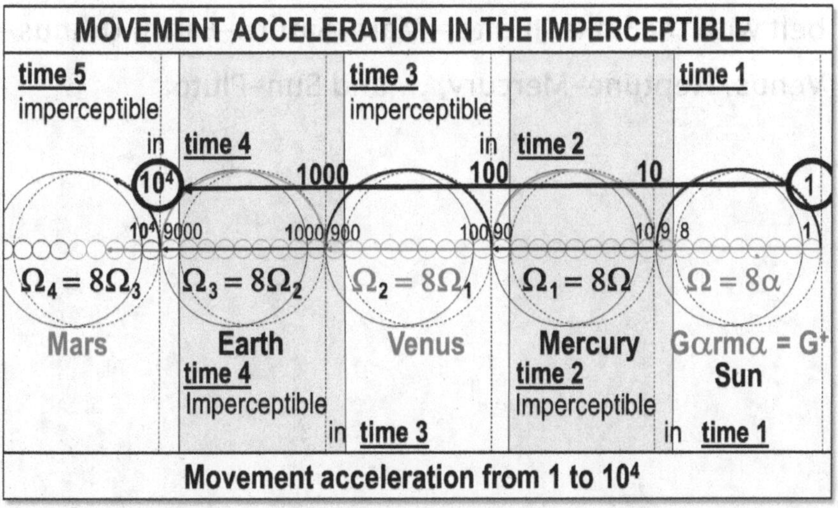

We recover the initial conditions of the 1st time at the end of the 4th time. The observer of time 4 on Earth thus has the time to make "the future" that the observer of time 1 wished in an instant in his time.

The Earth becomes the equivalent of the Sun for the time of Mars which becomes an initial time like the time of the "Gαrmα" horizon whose essential features we will see.

We will also see that the planets (Jupiter, Saturn, Uranus, Neptune and Pluto) correspond to the doubling movement between the Kuiper belt and the asteroid belt with the links: Jupiter–Mars, Saturn–Earth, Uranus–Venus, Neptune–Mercury, ... and Sun–Pluto:

Because the doubling movement proves us that, far from being a planetoid, as it's currently believed, **Pluto is an (α) planet linked to the Sun (Ω):**

Association of horizons and time openings

$i^{19} = i$	$i^{22} = 1$	$i^{25} = i$	$i^{28} = 1$	$i^{31} = -i$	$i^{34} = -1$	$i^{37} = i$
Asteroid)⁺	Jupiter	Saturn	Uranus	Neptune	Pluto	Kuiper's Belt
(Asteroid)⁻	Mars	Earth	Venus	Mercury	Gαrmα	Sun⁻
$i^{16} = 1$	$i^{13} = i$	$i^{10} = 1$	$i^{7} = i$	$i^{4} = -1$	$i^{1} = -i$	$i^{0} = 1$
				(Oort Cloud	horizon) =	Kαrmα $i^{40} = 1$

14 derivations = (×i) = 14 times at the same time

Beyond Pluto, the double Kuiper belt separates us from the Oort cloud which, for us observers, is surrounded by an "imaginary" horizon which I called **Kαrmα** ($i^{40} = 1$) because it is associated with the **Gαrmα** ($i^{1} = i$) horizon of our Sun. We shall see the reason for this association.

Between these doubling horizons, differentiating the times, it is therefore essential to have 3 information speeds linking the 3 simultaneous times: past (1), present (4), future (7).

But only the smallest C_0 will be observable in our time. The doubling movement which enables calculate them governs our solar system, wedged between the infinitely small of the particles and the infinitely large of the universe.

It is with these 3 doubling speeds that the planets of our solar system follow this fundamental movement.

The Earth ensures the passage of information between Venus and Mars, because we are the vectors of information circulating with a constant C_0 speed which is that of our light. Our Sun sends us vital information with this speed which, on the Earth, is the constant of our doubling.

The doubling theory enabled the speed of light to be calculated for the 1st time, while proving that, according to Einstein's paradox, it is indeed independent of the speed of the observer and the speed of the light source.

This paradox finally becomes explicable and makes us understand the principle of the relativity between the alpha and the omega which was well known to our distant ancestors.

4.3. The creation of the 1st solar horizon Gαrmα G+

It is the "2/3 law" which, as we have seen, requires the intermediate horizon Gαrmα (G+) between the Sun and Mercury.

The doubling movement enabled me to define this horizon which had been sought in vain *(in the 19th century by Urbain le Verrier who was looking for the invisible planet Vulcan)* before Einstein's theory of relativity.

Because of the existing laws, one imagined a planet between the Sun and Mercury.

The theory of relativity made it possible to exclude this possibility, which was therefore finally abandoned. In reality, it is the doubling of an "imaginary" horizon in the mathematical sense of the term, but real although unobservable except by its effects.

Indeed, the 2/3 law makes the planetary orbits oscillate around a multiple of 27 theoretical radial paths or 27 diameters ($D_S=\alpha/8$) of the Sun. For it imposes 9 radial paths, (external observation)$^+$, for 8 radial paths, (internal observation)$^-$:

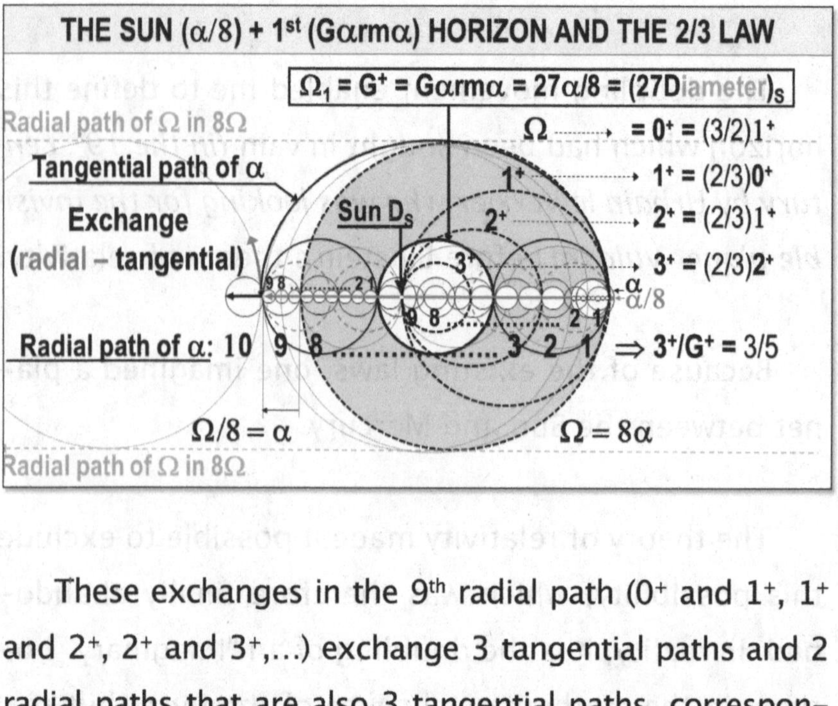

These exchanges in the 9th radial path (0^+ and 1^+, 1^+ and 2^+, 2^+ and 3^+,...) exchange 3 tangential paths and 2 radial paths that are also 3 tangential paths, corresponding to 2 radial paths that are also 3 tangential paths, etc..

According to the 2/3 law, the Sun is a α/8 doubling particle in a horizon whose radial path is equal to 27 solar diameters ($G^+ = 27D_S$) for the (internal)$^-$ observer and only 24 for the (external)$^+$ observer.

The (radial–tangential) exchange of the Sun uses a 9th time that is as imperceptible as it is explosive. With the planetary associations, whose importance we will see, **this law explains the current solar explosions** and shows us that they alternately concern the (radial–tangential) exchanges of the 12 planetary horizons:

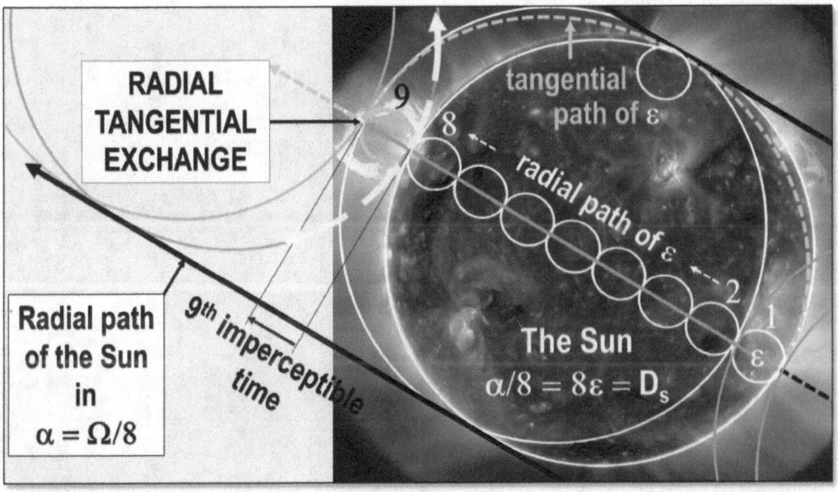

Thus, the tangential path of the α particle on the Ω = 8α horizon corresponds to the 8α radial path of the α particle in the Ω horizon, which, for an (external observer)+, moves radially by 1/8th of its diameter.

CHAPTER 05

THE DOUBLING EQUATION THE (Gαrmα)$^{+}$ HORIZON IN THE SOLAR SYSTEM

5.1. The largest structures in the universe rotate

"Some of the gigantic filaments that make up the "cosmic web" - an unfathomable network of tendril-like structures linking the universe's galaxies - seem to turn. These cosmic filaments, which are the largest known structures in the entire universe, rotate around their axis, according to researches published by Libeskind Noam."

(Nature Astronomy – 14th of June 2021):

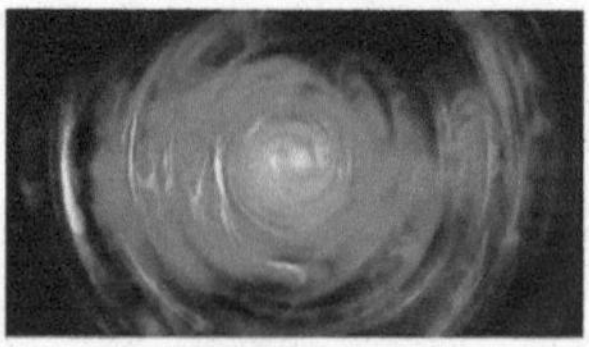

A cosmologist (from the Leibniz Institute for Astrophysics in Potsdam, Germany) states that scientists previously assumed that clusters of galaxies, which move very slowly around their axis, were the largest rotating objects in the cosmos. Rotation is common in space on smaller scales, but nothing in the Big Bang could have set the massive filaments in motion, given their incredible scale.

Only the current end of a doubling time cycle can explain it!

And this end disrupts our solar system.

The doubling movement in the infinitely small explains **the "spin" of particles** that turn on themselves as they move forward.

This end also concerns the infinitely large stars, galaxies and universes that surround us winding and unwinding on themselves. If the measure of our time changes, we can notice it by the horizons surrounding our own:

We can observe an expansion of the universe or a contraction, an acceleration or deceleration of this expansion which supposes a Big-Bang, or a Big-Crunch.

This observation is an illusion linked to the deceleration or acceleration of our observation time of a static universe.

In fact, a giant "roll-up" is cyclically made and unmade.

Thus, at the present end of a doubling cycle of time in our solar system, we are seeing sudden accelerations in the expansion of the universe before final contractions that will soon be just as sudden.

This will give us back the vision of an "initial horizon" and, as ancient texts say, we will have the illusion that *"the stars are falling from the sky"*.

But, like the Big Bang, this illusion will just only be the consequence of the end of a doubling of time cycle.

5.2. The doubling of time equation

A fundamental equation connects an (external observer)$^+$ who considers the Ω horizon as a particle in his 8Ω horizon and an (internal observer)$^-$ who considers Ω as the horizon of a ($\alpha = \Omega/8$) particle limiting his observation.

We are the (internal observers)$^-$ of our planet, small terrestrial particle in a solar space, considered in the galaxy as a simple star by the (external observers)$^+$ who evolve in other stars.

We are also the (external observers)$^+$ of the (internal observers)$^-$ evolving in their time. Seen from the outside, the $(\pi)^+$ rotation of Ω of radius R^+ corresponds to the $(2\pi R)^-$ rotation seen from the inside, while the radius $(R/8)^-$ seen from the inside becomes $(R)^+$ seen from the outside.

(Internal)$^-$ and (external)$^+$ observers are therefore linked by **the fundamental equation:**

$(\pi R^2)^- = (4\pi R)^+$.

THE DOUBLING EQUATION: $(\pi R^2)^- = (4\pi R)^+$

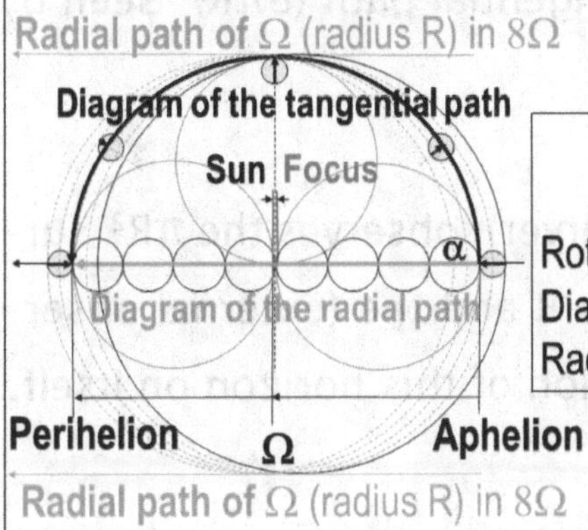

2 different observations :
(external)$^+$ and (internal)$^-$
Rotation: $(\pi)^+ = (2\pi R)^-$
Diameter: $(R)^+ = (R/8)^-$
Radial path: $(\pi \times R)^+ = (2\pi R \times R/8)^-$

Linking a surface to a tangential path around this surface, this equation clearly shows that the "radius-vector" (Sun-Planet) of a planetary orbit can sweep equal areas in equal times. It links the $(\pi R^2)^-$ surface of a Ω horizon of radius R to the $(4\pi R)^+$ radial path of this Ω horizon in the 8Ω dilated horizon.

The (Gαrmα)$^+$ horizon thus makes it possible to finally understand the reason for Kepler's 2nd law.

5.3. The (Gαrmα)$^+$ horizon and Kepler's law of areas

The doubling equation defines the radial path $(2\pi R^2)^-$ of a horizon of radius R^2 for an (internal observer)$^-$ with regard to the tangential path $(8\pi R)^+$ seen by an (external observer)$^+$.

Thus, the (internal observer)$^-$ observes the πR^2 surface of a horizon of radius R and the (external observer)$^+$ observes a $4\pi R$ rotation of this horizon on itself.

Moreover, this ray, connecting the Gαrmα horizon to the Sun, sweeps equal areas in equal times. It is the same for the Ω horizon.

This equation demonstrates Kepler's 2nd law relating a planet's travel time on its orbit to the surface swept by the solar ray of this orbit.

This ray sweeps equal areas in equal times

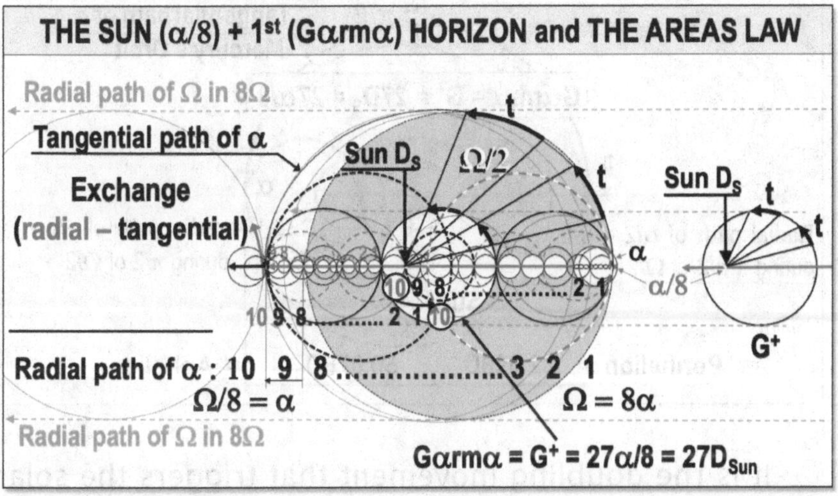

By following the radial path of the horizon or of the Gαrmα sphere (whose equatorial plane is G^+), we determine the position of the Sun ($\alpha/8 = D_S$).

Due to the $\pi/2$ rotation of $\Omega/2$ in Ω, which moves at the same time (radial path in the 8Ω horizon), the 8th radial path of $\alpha/2$ in $\Omega/2$ corresponds to the 1st in $\Omega/2$.

DIAGRAM OF THE RADIAL PATH OF THE G⁺ HORIZON

π of Ω/2 movement in Ω in π/8 movement in 8Ω ...

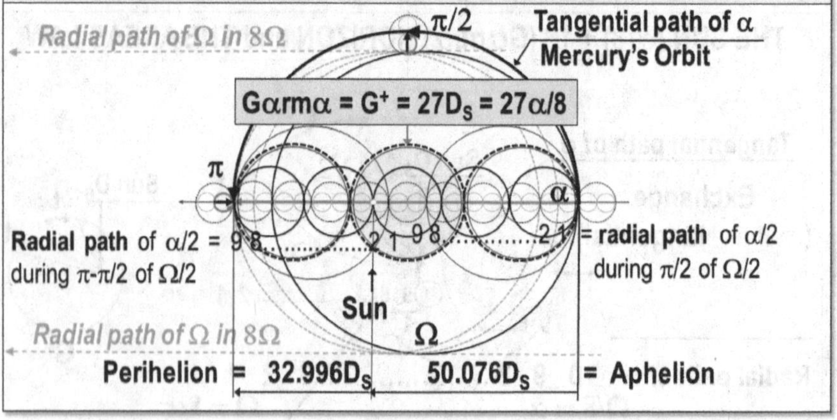

It is the doubling movement that triggers the solar explosions.

As there are 12 horizons in the doubling movement, there are 12 radial paths in 12 planes.

Thus, **Mercury's trajectory** (1st planet of our Sun) corresponds to the radial movement in the α tangential particle which performs its tangential path on the Ω horizon, itself in doubling movement in the 8Ω horizon.

5.4. The solar System

The doubling movement is such that the observers of the doubled spaces in different times evolve "at the same time", but in different times and planes.

We have seen that the "Gαrmα" horizon corresponded to the tangential horizon of the radial path of the Sun which, itself, was the 1st horizon. The 2nd horizon performs a 2nd doubling of time, with Mercury:

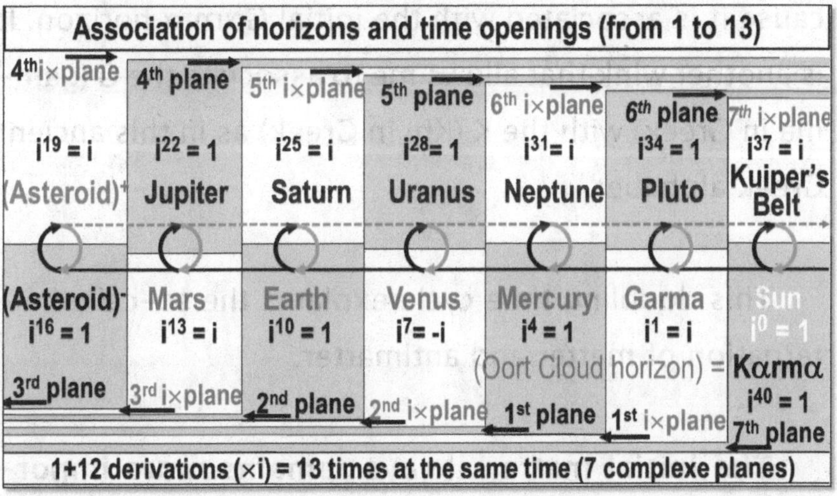

Mercury performs a 3rd with Venus[1]; which performs a 4th with the Earth; which performs a 5th with Mars; which

[1] Venus is in the $i^7(X, Y) = -i(X, Y)$ imaginary plane, associated with the (X, Y) initial plane in the [(X, Y) + i(X, Y)] complex plane. This explains the retrograde movement of Venus.

performs a 6th with the double Asteroids belt, of thickness $3G^+$ (9th radial path of $9x3G^+$).

This 1st part of the doubling time cycle, makes 6 doubled times work at the same time. Thus, the 2nd time hides the 1st one in the imperceptible which continues simultaneously, but in an accelerated time.

The time doubling movement thus uses seven complex planes. I have called Kαrmα the final horizon because it is associated with the initial Gαrmα horizon. It is another wink that allows me to associate the G (gamma in Greek) with the K (Khi in Greek) as in this ancient Greek alphabet.

This doubling time cycle explains the so-called alternation of matter and antimatter.

The initial matter always remains more important than the doubled matter (which is unjustly called antimatter) because the latter is, in its turn, a 2nd initial matter which always remains more important than its doubled matter in the imperceptible of the "observers" of this 2nd initial matter...

And so on, from the Gαrmα surrounding our Sun to the Kαrmα surrounding our solar system which is the horizon of our Sun for the "observers" of the initial matter...

5.5. The acceleration of times depends on "Gαrmα"

The 6th doubling is therefore that of the double asteroid belt (24 and 27)G+. The 24 G+ are three times 8G+ which give three more imperceptible 9th times:

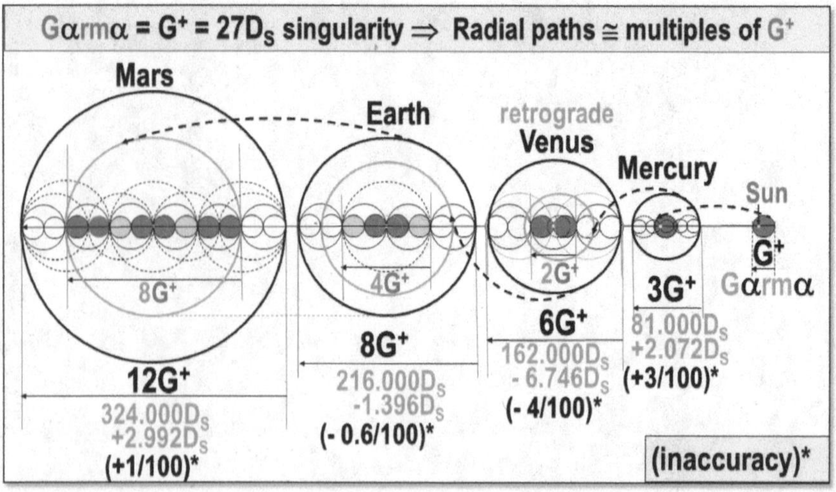

The movement continues with the giant planets:

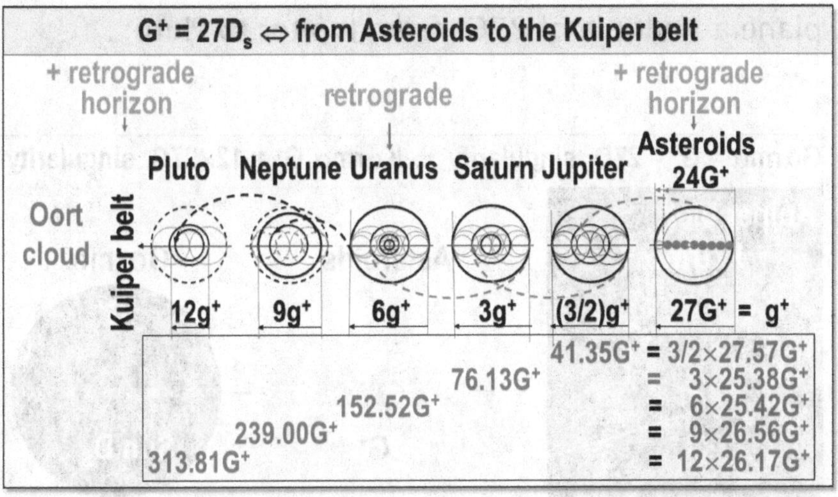

Showing the planes of the planetary orbits in a single plane, **the above diagrams are inaccurate.**

They do not take into account the radial rotation of the Sun in the "Gαrmα" horizon which makes a tangential path in the galaxy which, itself, makes a radial path in a galactic cluster which, etc...

Taking into account the inclinations of the orbital planes, in the plane of the asteroids, we obtain in this plane a multiple of 27G⁺ from Jupiter to Pluto:

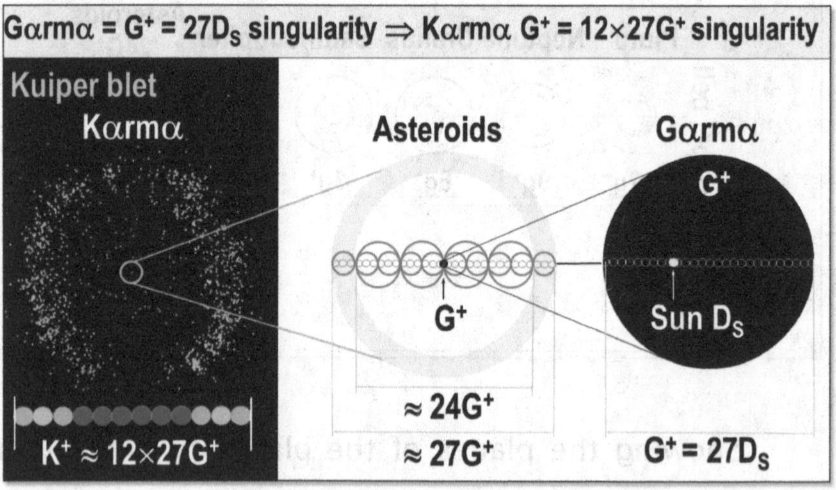

The movement adds "six doublings" up to the double Kuiper belt ($\approx 12\times 27G^+$) which borders the Oort cloud.

Thus, the Kαrmα K^+ horizon is associated with the Gαrmα G^+ horizon by the double intermediate link formed by the double Asteroids belt (24 y 27)G^+.

The doubling equation enabled demonstrate the link between the small planet Pluto (R_p = 1,180.75 km) which finds its place back in the solar system and the Sun ($2R_s = D_s$ = 1,394,180 km with the chromosphere), such that: $D_s = R_p^2$

We can therefore change the time scale by taking the square root of a distance that measures our observable space.

CHAPTER 06

THE 3 ENERGIES AND THE 3 DOUBLING SPEEDS

6.1. The three doubling energies

A doubling space is at the same time doubled by a 1st and doubled by a 3rd. Each of these 3 successive spaces must have the same radial coherence energy, which is 33.3% of the doubling energy.

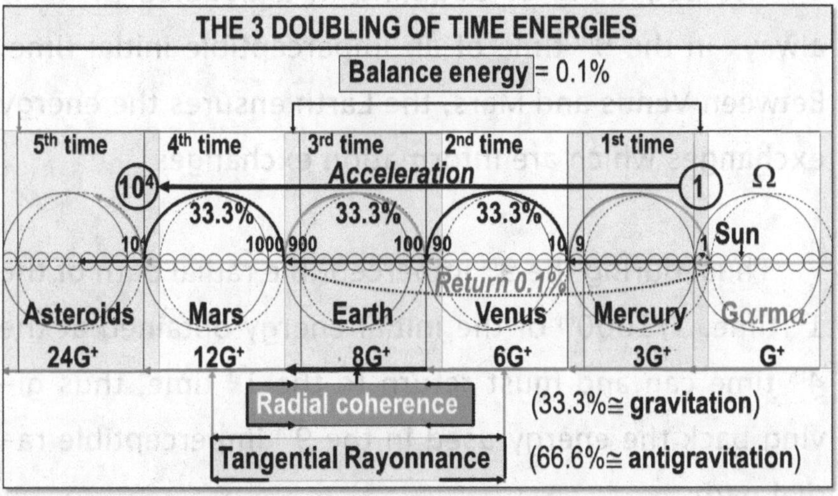

Since energy exchange take place in the imperceptible 9th radial paths, 33.3% may remain before the 1st path and 33.3% after the 8th one. For the observers that we are on Earth, we can say that:

- **66.6% is "the tangential rayonnance energy"** due to the exchanges in the imperceptible 9th radial paths.

- **33.3% is "the radial coherence energy".**

- **0.1% is "the energy of balance"** due to exchanges of information in imperceptible times that I have called **"time openings"**.

The beginning of doubling in a doubled plane is always in the 9th time of an imperceptible initial time. Between Venus and Mars, the Earth ensures the energy exchanges which are information exchanges.

Thus, during the 9th imperceptible radial path of the 1st time, 1/1000th of the initial energy obtained at the 4th time can and must return to the 1st time, thus giving back the energy used in the 9th imperceptible radial path.

The loss of energy necessary for this 1st doubling is imperceptible in the 1st time. There is therefore 999/1000th of this energy left at the end of the 9th radial path of the 3rd time, just before the 1st radial path of the 4th time.

Balance will be obtained and doubling will continue in the 4th time if this lost 1/1000th returns in this 1st time using the 9th radial imperceptible time of the 9th radial particle of the 3rd time. On the other hand, if this 1/1000th disappears in the 4th time, the information will not pass any more through the time openings to maintain the balance of the energies of rayonnance and coherence.

According to the previous diagram where the Earth is between the Sun and the asteroid belt, this means that we have to return this "thousandth" to the right in the past (fast time of the past = Gαrmα) in order not to be marked with the number 666 on the left (slow time of the future = asteroids and Kαrmα). This mark would indicate that our energy of rayonnance is only at the service of the future without any memory of the past. We would thus be in the greatest imbalance, far from all information and vital synthesis.

Indeed, in this case, our energy of coherence is transformed into an energy of rayonnance which makes us prisoners of the future without possibility of creative information of the past. It is thus that the horizon of our Earth becomes a Sun to light the darkness of the future.

The ancient writings spoke of the prisoners of the future (which would be marked with the number 666 of the future) and of those who direct it and whom we will discover at the end of the doubling of time... This has become the "beast of the apocalypse" from the Greek: the discovery (calypsos) of the hidden below (apo).

Current science speaks of dark energy, in the sense that this mysterious energy is only perceived through its effects. In reality, the mystery lies in the fact that we perceive only a small part of the doubling energy "in our time" which is not universal.

The universe which is perceptible in our time is not perceived the same way by (external observers)$^+$, who, evolving in an accelerated time compared to ours, seem to be in our past. Conversely, (internal observers)$^-$, evolving in a slowed down time, seem to be in our future.

In reality, the 3 speeds of information exchange between these three times give the impression of a before (C_2), corresponding to (external observers)$^+$ from the past, a during (C_0) which is ours, and an after (C_1), corresponding to (internal observers)$^-$ from the future:

And always such that: $C_0 < C_1 < C_2$.

With speeds greater than the speed of light which, as we have seen, are not perceptible in our time.

6.2. The 3 doubling speeds

The 3 successive horizons (Venus–Earth–Mars) concern us. We can define their observation time as "before" for Venus, "during" for the Earth and "after" for Mars.

Characterised by the speed of our light C_0, our time is in the imperceptible 9th radial time of Venus and the time of Mars is in the imperceptible 9th radial time of the Earth.

The doubling movement of time imposes an outward speed C_1, a return speed C_2 greater in the imperceptible time of the 9th radial path of the 3rd time.

These 2 speeds must be greater than the speed C_0 of the radial path of the 4th time, and such that:

$C_2 > C_1 > C_0$

The speed of "our" light C_0 is the only speed of the observable information in our present time that concerns the space of the Earth.

The speeds C_1 and C_2 are information speeds for

(internal)⁻ or (external)⁺ observers that the doubling movement enable us calculate, but which, by definition, remain unobservable.

Indeed, these information circulates in the 9th imperceptible time of the 4th time (ours) observer who can therefore not observe them.

The passage of information between doubled times (1st, 4th, 7th) requires common imperceptible times between the 1st time that our 4th time doubles and the 7th time that doubles our 4th time.

The information uses the imperceptible "time openings" of the 9th radial paths between these three embedded times in the same doubling movement. Our Sun is, for us, the source of the vital "luminous" information.

This information only depends on the doubling movement. It does not depend at all on the speed of the observers that we are on our planet nor on the speed of our Sun, source of our light.

It is therefore not a speed in the strict sense of the word.

We find Einstein's paradox, explaining that light always catches up with us at the same speed because this constant is independent of the speed of the observer and the speed of the light source!

In an observation horizon that our planet offers us, it is normal that each observer can receive the same information at the same time. This is the only way we can all evolve at the same time with the same vital information arriving in perfectly defined imperceptible times.

"Quantum" mechanics had already noted this without explaining it, because the "quanta" of energy arrive in unobservable times but which are not null: this is Heisenberg's Law.

Doubling shows that an imperceptible time, for the observers of our time on our planet, can become a real time for an observer of another world, doubled from ours, evolving in another time. Likewise, an observable time is an imperceptible time for an observer evolving in a time of which ours is doubled.

The measure of our time is calculated by the rotation of the Earth around the Sun (a year) and by the rotation of the Earth on itself (the day):

There are 365.256 363 days in a year.

A day is made of 23h 56m 4.0845s

Defined by Kepler, the quasi-elliptical movement of the planetary orbits is in reality only an approximation of the doubling movement. It requires a large correction:

It is the famous "**Keplerian anomaly**"!

We will see that the doubling movement makes it useless while explaining it.

Our time is perfectly defined by the orbit of our planet. Its radial path goes from the furthest point from the Sun (aphelion) to the closest point (perihelion):

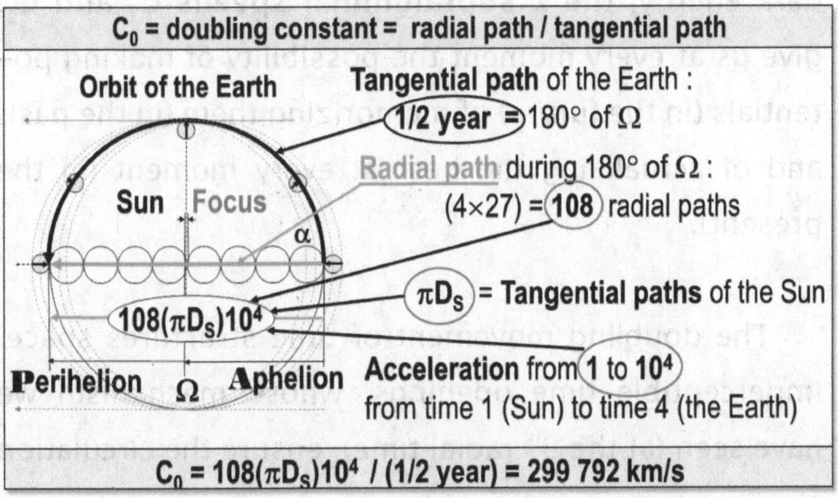

We have seen that with an acceleration of the movement from 1 to 10^4, this radial path corresponds to:

$4 \times 27 = 108$ tangential paths of the Sun πD_S,

D_s = 1,394,180 km is the diameter of the Sun that corresponds to the radial path of the Earth and the Moon on the Ω horizon (from the aphelion to the perihelion). This radial path lasts ½ year.

However, it is the ratio of the radial path 108 $(\pi Ds)10^4$ and the tangential path (½ year) that defines the C_0 doubling movement constant:

$$C_0 = 108(\pi Ds)10^4 \; / \; (½\ \text{year}) = 299{,}792\ \text{km/s}$$

Not yet accepted and as poorly understood as the dark energy, **the 2 supraluminal speeds C_1 and C_2**, give us at every moment the possibility of making potentials (in the future) of memorizing them (in the past) and of actualizing the best at every moment (in the present).

The doubling movement of time structures space. Imperceptible time openings, whose mechanism we have seen (at the 9th radial time), ensure the circulation of vital information between the three doubled times (past – present – future) with the three speeds C_2, C_0, C_1, such as: $C_2 > C_1 > C_0$.

This relation is deduced from the doubling movement. Indeed, an information which leaves from an observer of the Earth (time 4) in an imperceptible 9th radial time has a speed C_1 greater than C_0.

For all our questions must reach all our (internal observers)⁻ before receiving the answers from our (external observers)⁺.

This speed C_1 must therefore be a multiple of both 4 and 3. Indeed, according to the "2/3 law", seen earlier, the (internal)⁻ observer perceives the radial paths as multiples of 4 and the (external)⁺ observer as multiples of 3. This requires that $(3\times4)C_1 = 12\ C_1$ be a multiple of C_0.

This C_1 speed of information leaving our time is accelerated from 1 to 10^5 compared to the information arriving to us at C_0 speed. It is also necessary that $12C_1$ be a multiple of 10^5C_0.

Now, the information arrives in our time, at the 1st

radial path, when the information leaves our time at the 8th radial path in an imperceptible time. It must therefore be 7 times faster and such that:

$$C_1 = (7/12)10^5\ C_0$$

It is also necessary that the information of "time 5" returns to "time 0", in order to be controllable by the (external observers)$^+$ of which we are the (internal observers)$^-$ of time 4, i.e. :

$$C_2 = 7C_1$$

Conclusion:

$$C_2 = 7C_1 = (7^2/12)10^5\ C_0$$

Thus, the observers, whom we are in our time on our planet, have the answer at the same time as the question.

It is the imperceptible times, defined by the doubling movement, that give the impression of a miraculous entanglement between doubled observers (or particles) in different times.

These three information speeds allow us to understand the doubling of time movement that we are "wrongly" using in our present.

Ignoring this **vital mechanism**, we have become slaves to the (internal observers)$^{-}$ who have time to experience our future potentials.

By distancing us from the (external observers)$^{+}$ who make the best synthesis at each moment, the latter direct us by pretending to be gods, creators of our future. But they are only our servants because our thoughts are their energy!

To get out of this trap, all we have to do is listen the (external observers)$^{+}$ and, to do this, we must not ignore the mechanism of supraluminal information.

We know that the presence of an observer can modify an experience. It is therefore logical to think that an observer modifies the future and, therefore, the actualisation of a future potential in the present.

With its Gαrmα "gearbox", our solar system ensures these anticipatory information exchanges: talking about "entanglement" between particles does not explain this rigorous mechanism of time doubling.

These 3 speeds are used by the information that passes from one time to another in the 9th imperceptible times. They are therefore independent of the speed of the observers that we are on Earth and of the (internal)$^-$ or (external)$^+$ observers.

They suppress the Einstein paradox and the impossibility of observing in our time "the time openings", sometimes called "wormholes" in order to justify supraluminal speeds inconceivable, except by the doubling of time theory.

Thus, at each moment, these “time openings” provide us with the best and worst potential states according to the analysis that our present imposes on us.

The calculation of these three speeds makes it possible to better understand the mechanism of information exchanges between doubled times.

During our REM sleep, our energetic body uses these three speeds to free itself from our physical body and sends us back vital information: if we suppress this vital sleep, we will die after 3 weeks! This is a perfectly proven law, but it remains unknown.

With the doubling of time, the explanation is simple: from time 1 (the Sun) to time 4 (the Earth), we dispose of an acceleration of time from 1 to 10^4 (previous figures).

The half-turn or rotation of 180° of the Earth around the Sun, corresponds, according to the 2/3 law, to (3×2) = 6 radial paths, that is ½ year of our calendar:

$$1 \text{ year} \approx 2\times180° \approx 365.25 \text{ days} \approx 12\times30.44 \text{ days}.$$

CHAPTER 07

THE PLANETARY ORBITS OF THE SOLAR SYSTEM

7.1. Link between the planets of our Sun

Regardless of all the more or less logical hypotheses formulated over the past centuries and millennia, the doubling equation rigorously shows the link between each small planet (known as telluric) and a giant planet (known as gaseous). Particle associated with the Sun, Pluto regains its title of planet.

This association is as follows:

Sun	Pluto
Mercury	Neptune
Venus	Uranus
Earth	Saturn
Mars	Jupiter
The Gαrmα horizon	the asteroids belt

The **Gαrmα** horizon corresponds to ($G^+=27D_S$). As we have seen, the asteroid belt ($27G^+=27^2D_S$) is therefore linked to the **Kαrmα** ($12\times27G^+$) horizon which is itself linked to the Kuiper belt.

7.2. The planet Mercury

With the radius of Mercury R_{Me} = 2,439.90 km and a rotation of 180°⇔ π, **the doubling equation gives the distance of the foci of Mercury's orbit:**

$$(R_{Me})^2 = 17.080D_S$$

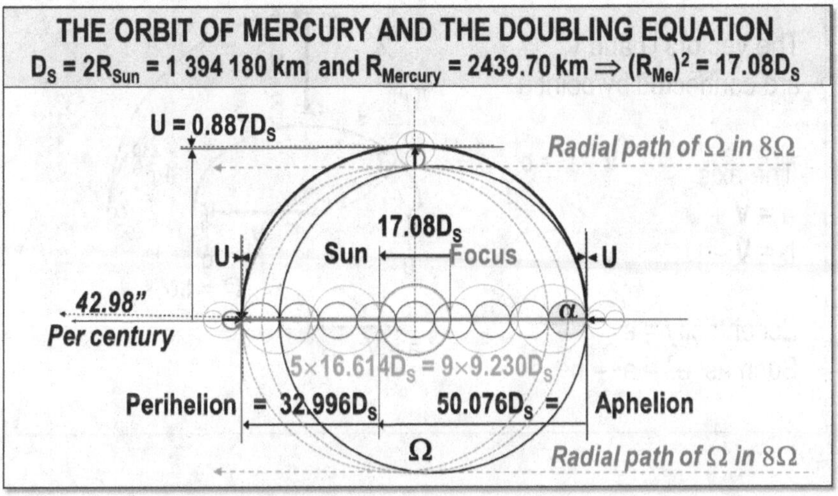

Because of the π rotation of Ω which moves at the same time (radial path in 8Ω), the 8th radial path of α in Ω corresponds to the 1st tangential path of α around Ω.

Thus, the trajectory of Mercury, the 1st planet of our solar system, corresponds to the radial movement "in" the α tangential particle which carries out its tangential path on the Ω horizon, itself in doubling movement in the 8Ω, horizon.

The doubling movement would correspond to the definition of an ellipse for a planetary orbit, if the radial path were independent of the tangential path:

The geometric definition of an ellipse in a stationary plane

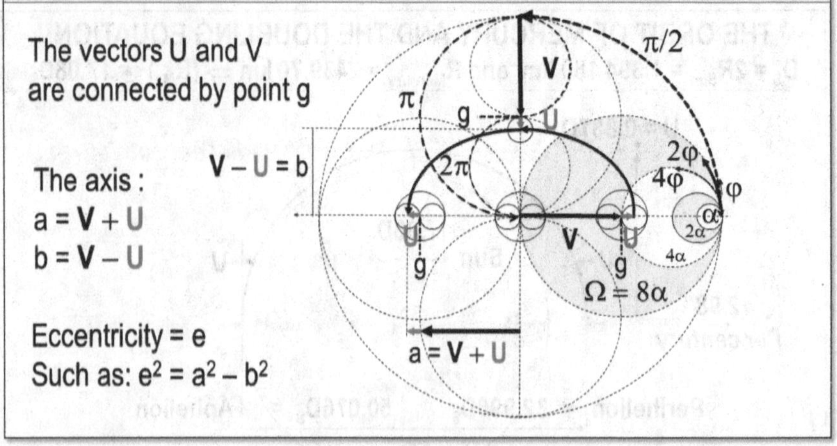

When the Ω horizon of Mercury makes a ½ turn π the α particle makes 9 radial paths in α due to the Ω radial path of Ω in 8Ω. Thus, the 10th radial path of α in Ω becomes the 1st in the 2nd of Ω in 8Ω:

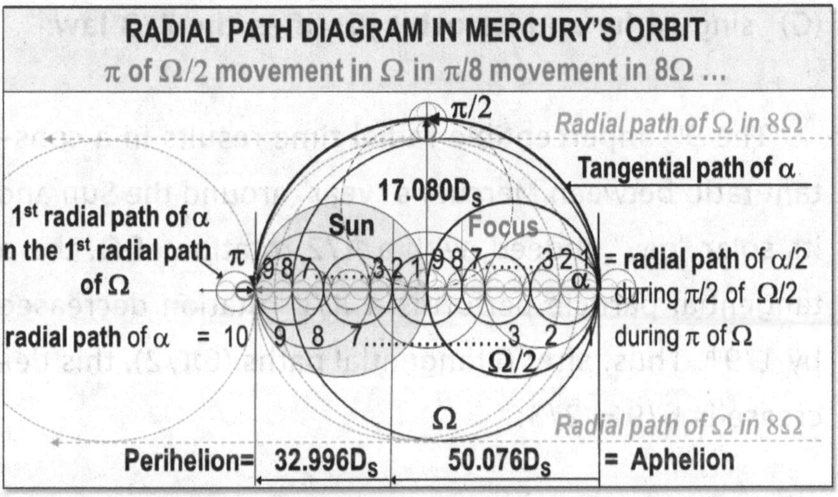

7.3. The 2/3 law rules Mercury

This law is verified with the Gαrmα horizon and the orbits of the planets of our Sun of which twinning verifies the doubling of time. Twinned with Neptune, Mercury is the closest planet to the Sun and to the Gαrmα (G)$^+$ singularity **and its orbit justifies the 2/3 law:**

The 9th imperceptible radial time results in a constant ratio between Mercury's "year" around the Sun and its solar "day". Indeed, with a $\pi/2$ rotation of Ω, the α tangential particle performs a $\pi/2$ rotation decreased by 1/9th. Thus, after 6 tangential paths ($6\pi/2$), this decrease is 6/9 = 2/3:

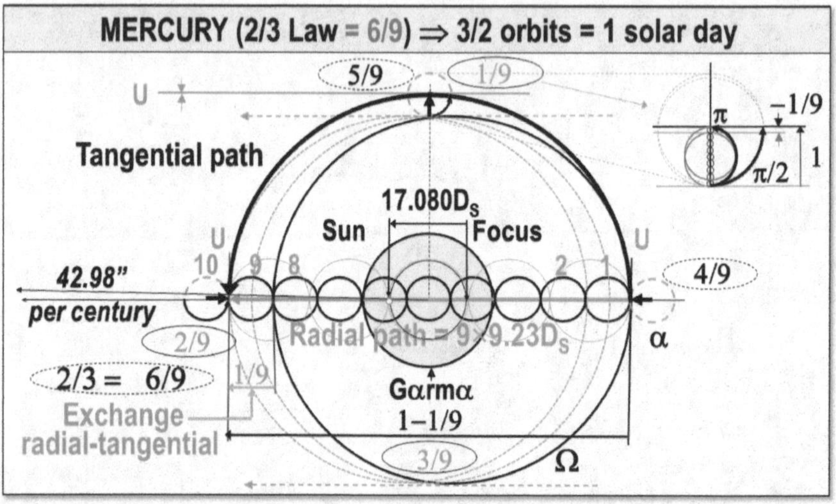

The movement clearly shows that the sidereal revolution period of Mercury (year of 87.969 terrestrial days) is equal to 3/2 times its sidereal rotation period: one Mercury day = 58.646 terrestrial days.

7.4. Precession of the axis of the Mercury orbit

Doubling implies the precession of the axis (aphe-lion–perihelion) of Mercury.

We have seen (chapter: 3.6.) that in the (X, Y) plane of time 1, the $\pi/8$ rotation ending the π rotation co-rresponds to the (X, Y) plane of time 4 and the $\pi/64$ final rotation to the (X, Y) plane of time 7.

With the acceleration from 1 to 10, from 10 to 100, etc... we obtain

$$(\pi/64)(1/10 + 1/100 + \ldots) \approx 0.309375° \text{ per cycle.}$$

We will see that this cycle is equal to 25,920 years.

We therefore obtain a precession of 42.98" per century.

7.5. Associated with Mercury: the planet Neptune

The planet Mercury is associated with the giant planet Neptune. The diagrams concerning the orbits of the planets are inaccurate because the planes and axes of the associated planets (like Mercury–Neptune) are not superimposed. However, with this superposition, they show that the doubling movement associates a telluric planet with a giant planet:

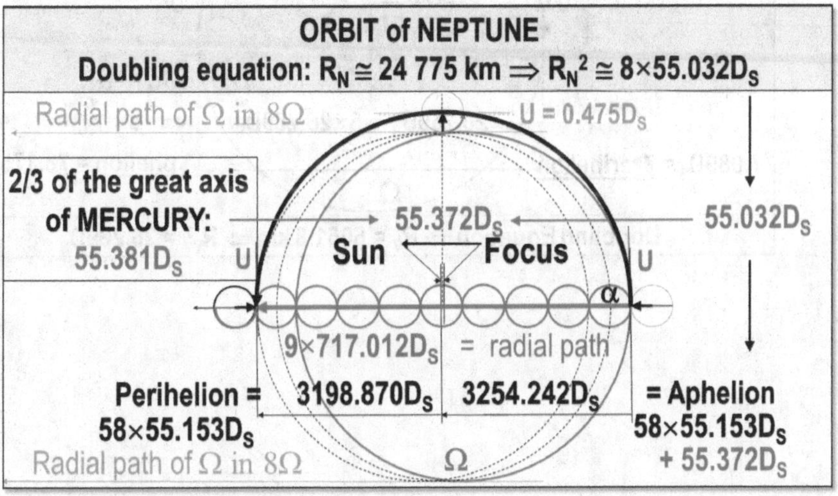

As expected, at the end of the doubling cycle, Neptune returns energy to Mercury.

7.6. The planet Venus

The doubling movement continues in the same way with the planet Venus which performs a retrograde movement due to the spin back of the plane of the orbit in the initial plane:

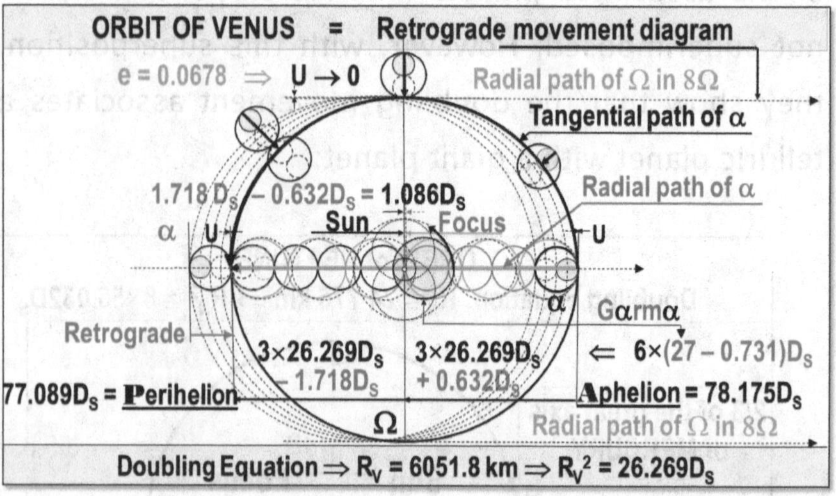

7.7. Linked to Venus: the planet Uranus

The movement associates the planet Venus with the planet Uranus, which returns energy to Venus at the end of the doubling cycle. As for Mercury and Neptune, the planes and axes of Venus and Uranus are not superimposed:

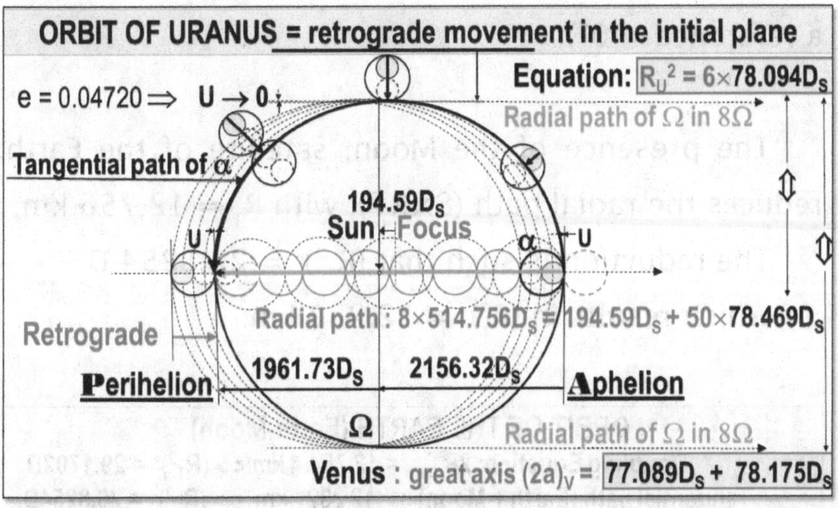

As predicted by the doubling movement, the planets Venus and Uranus are linked and both retrograde.

7.8. The planet Earth

The Earth is a telluric planet which is associated with the giant planet Saturn which returns energy to the Earth at the end of the doubling cycle. The planes and axes of the Earth and Saturn are not superimposed. The Saturnine plane is tilted by 2.486° with respect to the plane of the Earth (ecliptic which serves as a reference = 0°).

The presence of the Moon, satellite of the Earth, reduces the radial path ($8\times R_T^2$), with R_T = 12,756 km.

The reduction is such that $(R_T^2)^-$ = 26.8254 D_S

Which results in $(R_T)^-$ = 12,232 km:

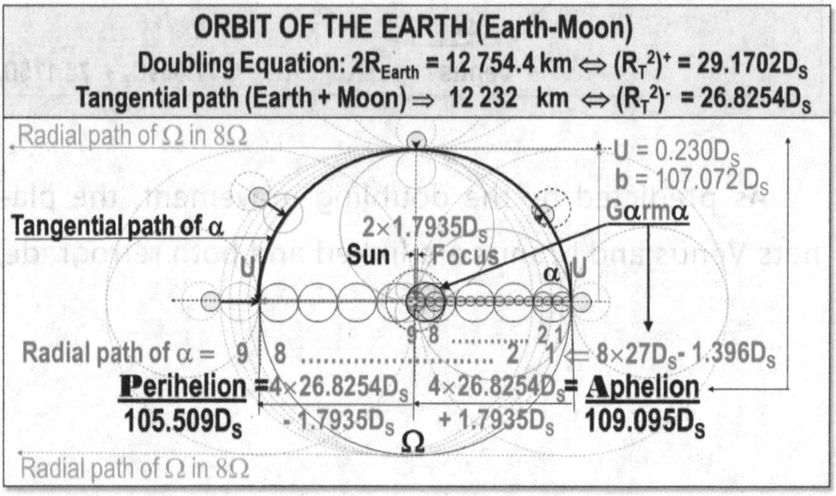

7.9. Linked to the Earth: the planet Saturn

The movement also associates the Earth with the planet Saturn, which returns energy to our planet at the end of the doubling cycle:

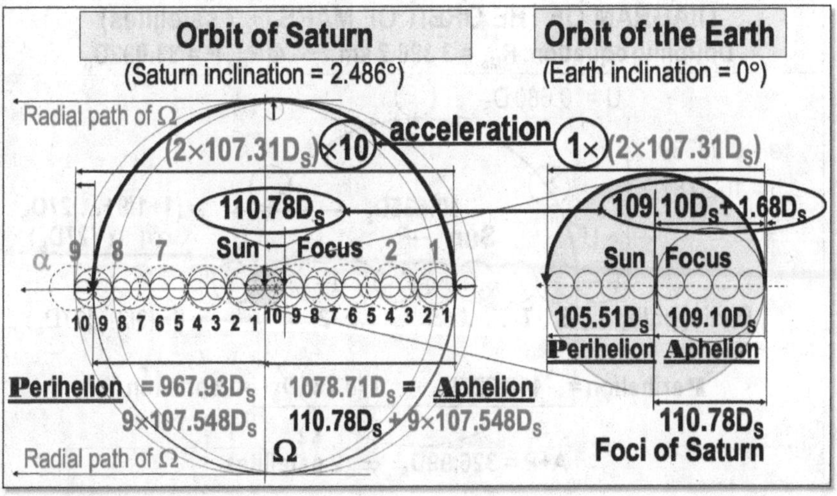

7.10. The planet Mars

Mars is a telluric planet associated with the giant planet Jupiter.

The presence of 2 satellites changes the theoretical radial path given by the doubling equation:

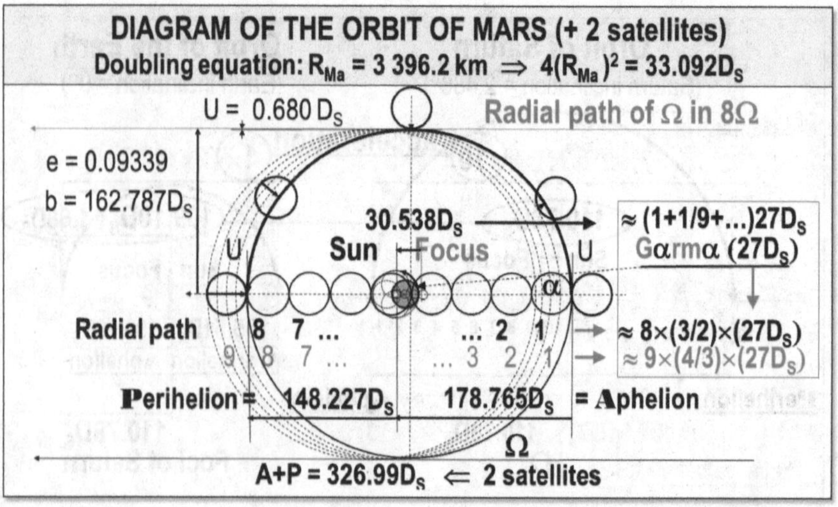

7.11. Linked to Mars: the planet Jupiter

The movement also associates Mars with the planet Jupiter which returns energy to Mars at the end of the doubling cycle:

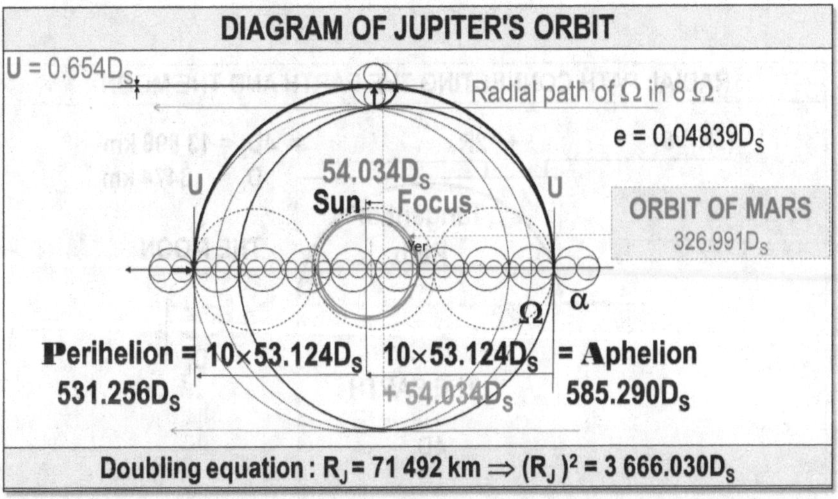

As with Mercury-Neptune, Venus-Uranus, Earth and Saturn, the planes and axes of Mars-Jupiter are not superimposed.

7.12 Association Earth-Moon

The 9th imperceptible time of the Earth's radial path gives the diameter of the Moon which, due to the permanent juxtaposition of the radial paths, always presents the same face to the Earth according to the following diagram:

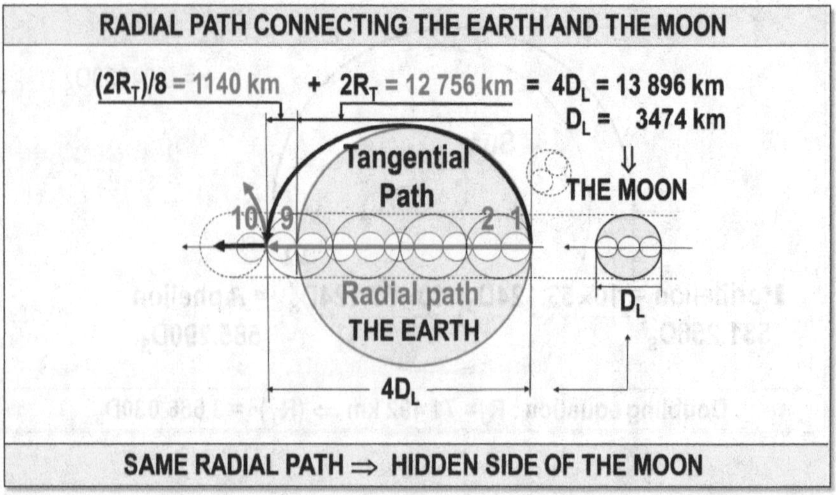

7.13. Association Sun-Pluto and Gαrmα-Kαrmα:

The doubling equation also gives us the diameter D_S of the Sun which is equal to the square of the radius of Pluto R_P^2. With a major axis $\approx 100\pi G^+$, Pluto is the planet associated to the Sun. Hence the following diagram:

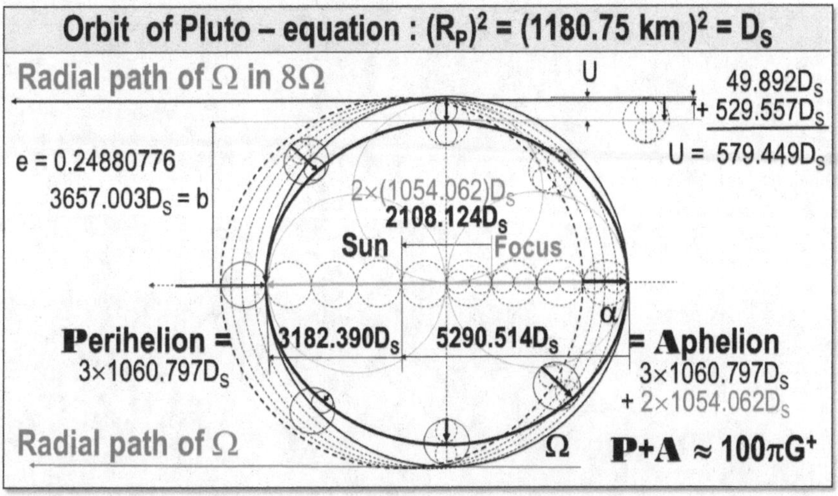

The doubling movement, which associates Pluto (α) with the Sun (Ω), also associates the Gαrmα ($G^+=27D_S$) "particle" with the Kαrmα ($K^+=12\times27G^+$) horizon, source of energy. This explains the fact that the giant planets give more energy than they receive from the Sun.

At the end of the doubling cycle, the contraction of the Kαrmα horizon seems to modify the orbits of these giant planets, while giving the impression of an acceleration of the expansion of the observable universe.

CHAPTER 08

THE TWO SOLAR CYCLES OF THE DOUBLING OF TIME

8.1. The precession cycle of the equinoxes.

This cycle concerns the observers that we are on Earth in our "present" time that we think universal, between the (internal observers)$^{-}$ of the "future" and the (external observers)$^{+}$ of the "past".

For ancient civilizations, it was the 25,920 years cycle of precession of the equinoxes, which includes a transition of 1,080 years (or 12×90 years) between two cycles of 24,840 years (or 12×2070 years).

These 12 periods correspond to ten years (or ten orbits) of Pluto.

We have seen that after six accelerations and dilatations, the six double spaces of our solar system double and accelerate six times, before returning to their initial position after six decelerations and contractions of up to maximum 30 years each.

We are currently experiencing the penultimate of these six 30-year periods, all of which begin with a huge solar explosion. The first 4 non-standard solar explosions that separated 3 periods of 30 years, took place in 1899, 1929, 1959, 1989, to the day, thus opening the doors to all potentials.

The 5th occurred in 2003, that is, 16 years too early, opening the doors to the past in advance, probably to return a lost balance between the past and the future! There are two more to come at the latest in 2033 and 2063.

Einstein had calculated a cosmological constant, thus supporting the idea of energy sharing, as in the doubling of time (33.3% and 66.6%).

Certainly, in the last century, no one imagined that the end of a cycle revealed, in six times 30 years, the end of the six doubling times of space and, above all, of time!

It was a very dangerous end that saw the sudden disappearance of Neanderthal man, 25,000 years ago!

But it was a beginning for the Cro-Magnon man that we are, with a small potential of Neanderthal (8% of our genome), which makes us wonder which extremely competent scientists could have given us this DNA 25,000 years ago!

Justifying the doubling of time theory, the 1st solar explosion in 1899, made us discover an expansion of the universe. The Big Bang seemed to be a good explanation.

The 2nd one in 1929 showed that this expansion was preceded by an inflation (acceleration of the expansion), which justified the Big-Bang hypothesis.

Forced to admit this initial explosion, Einstein would admit that his cosmological constant was "*the biggest mistake of his life*", even though it was a very big and very accurate intuition.

Finally, we discover the fossil radiation of the universe which seems to explain the initial energy: this is just a misinterpretation of the observation.

By ignoring the doubling of time, we ignore the change in time observation at the end of a doubling of time cycle which shows us an evolution of 25,920 years (i.e., 12 times 2,160 years) over a very short period of maximum six times 30 years!

After the 5th solar explosion of 2003, we still have to discover the last potentials in maximum 2 times 30 years. We will then rediscover our universe, as it was at the beginning of the last doubling cycle.

We are now living with all the potentials that humanity has made in imperceptible times. But we no longer know how to modify them before we live them.

We have not yet grasped the importance of this end of the doubling of time, which gives us the enormous possibility of living, at every moment, the best of the existing future potentials. There is still time to react with full knowledge of the facts to calm the planet because ignorance makes us live badly by making us accept and actualise in our present useless, dangerous or deadly potential futures, such as a surprising coronavirus or fratricidal wars...

With the knowledge of time doubling, we can be master of our future and live a present acceptable to all.

8.2. The Milankovitch cycle

Discovered by Milankovitch, this cycle is really about the (external observers)+ who are in another time. Our planet records the various cataclysms linked to this cycle.

It is the 2/3 law that separates us.

This cycle is thus 2/3 longer than the cycle of precession of the equinoxes (or Pluto cycle) which regulates our incarnation. Inexplicable or too badly explained, it corresponds to: 24,920(1+ 2/3) = 41,400 years.

It is thus linked to the cycle of precession of the equinoxes which, as we shall see, explains astrology, exact and ancient science that has been much abused and too often falsified by centuries of obscurantism.

8.3. Previsions concerning the solar system

With the doubling theory, we can finally understand why our giant planets move slightly in their orbit!

Beyond the Kuiper belt that surrounds our planets, the Oort cloud contracts (hence the arrival of planetoids in this belt). This gives the observers that we are, the impression of a lightning acceleration of the expansion of the universe.

After this acceleration of the expansion, we will soon observe such an accelerated contraction that we will feel as if "the stars are falling from the sky".

We are beginning to understand that "black holes" hide their reality by a lightning acceleration of time: a second can become billions of billions of seconds! The exchange of the radial and the tangential in a black hole at the 9th imperceptible radial path is a "singularity" corresponding to the famous proof by nine (9 = 0) in the calculations! At the end of a doubling cycle, the dilation of the horizons appears as so much singularity.

Black holes appear:

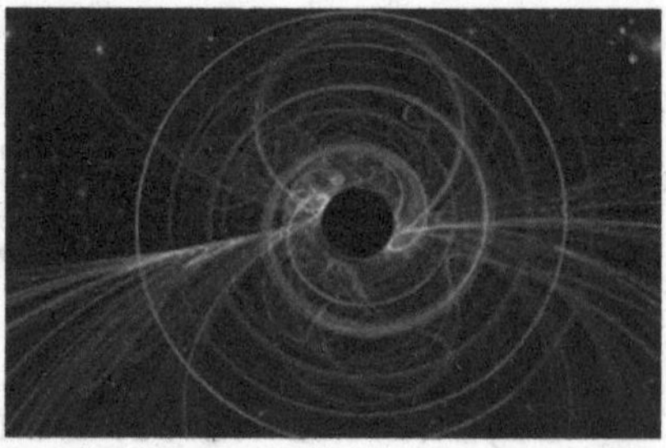

Black hole and Hawking radiation (Artist's drawing)

Highlighted by Stephen Hawking, and poorly received by the scientific community of the time, the radiation emitted by a black hole corresponds to the exchange of the radial and the tangential:

In April 2019, the collaboration of several scientists (Event Horizon) finally made it possible to get the silhouette of a black hole in the galaxy M87 with a jet of matter over more than 5,000 light years.

Anticipated by the doubling theory, as with the gravitational waves that have just been observed by the fusion of two black holes, this reality cannot be disputed.

This is explained by the doubling theory, which imposes doubling cycles of time of 25,920 years with an "apocalyptic" beginning and end.

In 1998, Saul Perlmutter and Brian Schmidt (Nobel Prize 2011) showed that the expansion of the universe seemed to be accelerating. Only the time doubling theory explains this acceleration which, at the end of the cycle, precedes the contraction.

The separation of the 12 acceleration and dilation times of the 12 observation horizons of our solar system that lasts 25,920 years fades in 7 times 30 years, giving a false impression of dilations and then lightning contractions of the universe that surrounds the solar horizon of our observation.

To achieve this balance, it is essential to know how to manage and control the vital balance energy (0.1%) which, between the 33.3% of coherence and the 66.6% of rayonnance, stabilizes our solar system. But we ignore its existence!

We don't even know anymore that our ancestors knew how to use it to try to get the best present condition... or the worst, because knowing a law makes it possible to skip it! It is urgent to take care of it and to master the energy of our doubling to survive the potential chaos.

Knowing where, when and how to give back the $1/1000^{th}$ of the energy we receive to live better... becomes essential. How much nonsense has been said and repeated concerning the 666 of which the ancient texts speak.

This 66.6% concern **the energy of coherence** and link us to the (internal observers)$^{-}$ who have the time to create our potentials at each moment.

At the present **end** of the current doubling of **time** cycle, we have this, $1/1000^{th}$ which we must give to the (external observers)$^{+}$ who have the best synthesis of our potentials.

In order to do this, we must know how to free ourselves from the (internal observers)$^{-}$, without inventing laws or dogmas, which appear to be logical but which, as the ancient Greeks taught, are nothing but pointless sophisms such as:

"*A cheap horse is rare. All that is rare is expensive. So a cheap horse is expensive!*"

Sophisms which, making the glory of current spiritual techniques, lead us towards unbearable dictatorships.

In the past, long before Christ, scientists, who were considered to be the "doctors of the law" of times, spoke about their knowledge only among themselves. They considered that not everyone could access their science. They reserved it only for those who were capable of understanding it, in order to avoid deviations.

Their knowledge was therefore reserved for certain initiates and therefore considered "esoteric". Esotericism was a precaution against possible and undesirable deviations. The time doubling theory can explain and rectify deviations without us being obliged to reserve it for a few privileged initiates who, as always, would take advantage of it to manipulate the ignorant.

CONCLUSION

Far from all postulates and beliefs where logic is often absent, it would be good, not to say urgent, to tame this law of time.

Governed throughout the universe by the doubling movement, it can, if we ignore it, lead us to an existence imposed, in the good or bad sense, by the future where (internal observers)$^{-}$ experience our potentials, in a time that does not have the time to appear in ours.

Taking advantage of the time openings that are nowadays permanent - due to **the** current **end** of **the time** doubling cycle - we could, on the contrary, live the best of futures.

It is enough to know that each of our thoughts produces potential futures (*in a time that does not have the time to exist in our time*).

These potentials are synthesised (*in a time where ours does not have time to exist*) before being suggested to us to live them by actualising them at each moment (*in our time*).

We can therefore anticipate the best way to live on a daily basis without wrongly believing, that a chance, happy or unhappy, guides us permanently into the unpredictable.

For, in ignorance of this law of time governed by the universal doubling movement of space and time, we would be under the permanent sway of servants of the future who know how to channel the whole world with skill, into the corridors of fear, doubt and their will.

The world must wake up before the end of the current cycle of time doubling. The universe will batter us without immediate awareness.

It is certainly not a question of being or becoming a bird of good or bad omen.

However, it is good to know, in order to live and survive well, that time doubling gives us the possibility to anticipate our present 2/3 of a second before living it and that every 8 minutes, every 3 days and every 40 days, we receive a synthesis of the future allowing us to anticipate and actualize the best potential in our time.

Highlighted by **Benjamin Libet**[2], these 2/3 of a second explain how the doubling particles are "entangled" in a mysterious way and receive the same information at the same time.

The time doubling theory explains how, but above all why!

For we too are entangled with a "double" whose existence we are unaware of and who, like an attentive but imperceptible "father", can suggest to us the best synthesis of our potentials at any moment.

[2]In 2003, the "virtual Nobel Prize in Psychology" was invented for Benjamin Libet's results in the field of consciousness, action initiation and free will.

The very recent discovery (November 2021) of ghost[3] neutrinos will certainly clarify the mechanism of anticipatory information in time openings as well as the mysterious quantum entanglement between two photons, highlighted by Alain Aspect which reveals the limit of the C_0 speed of information in our time.

The time doubling theory adds 2 information speeds C_1 and C_2, such that $C_0 < C_1 < C_2$.

Known as "supraluminal", these two speeds are undetectable in our time because they only concern information circulating in the imperceptible 9th time.

[3] In November 2021, physicists detected (Faser experiment) ghost particles in the Large Hadron Collider. These are signals indicative of neutrinos produced during particle collisions that can help scientists better understand the information in time openings.

But they are finally able to explain the mysterious entanglement of particles as well as our possible entanglement with what we could call a "double".

I am very grateful to the scientists who took the time to study and verify most of my findings and publied them in their congresses:

NOTE

NOTE

www.ingramcontent.com/pod-product-compliance
Lightning Source LLC
LaVergne TN
LVHW041208061126
851857LV00018B/1984
* 9 7 8 8 4 9 4 2 1 8 1 2 5 *